ISBN: 9781520154725
Imprint: Independently published
Copyright © 2016 Ela Joseph
All rights reserved.

You and I are made of cells

Written by Ela Joseph

Illustrated by MartinMak

To my Curious kids, who want to know everything

You and I

are made of cells

The plants and trees

are made of cells

Every living thing

is made of cells

Some are made of one cell

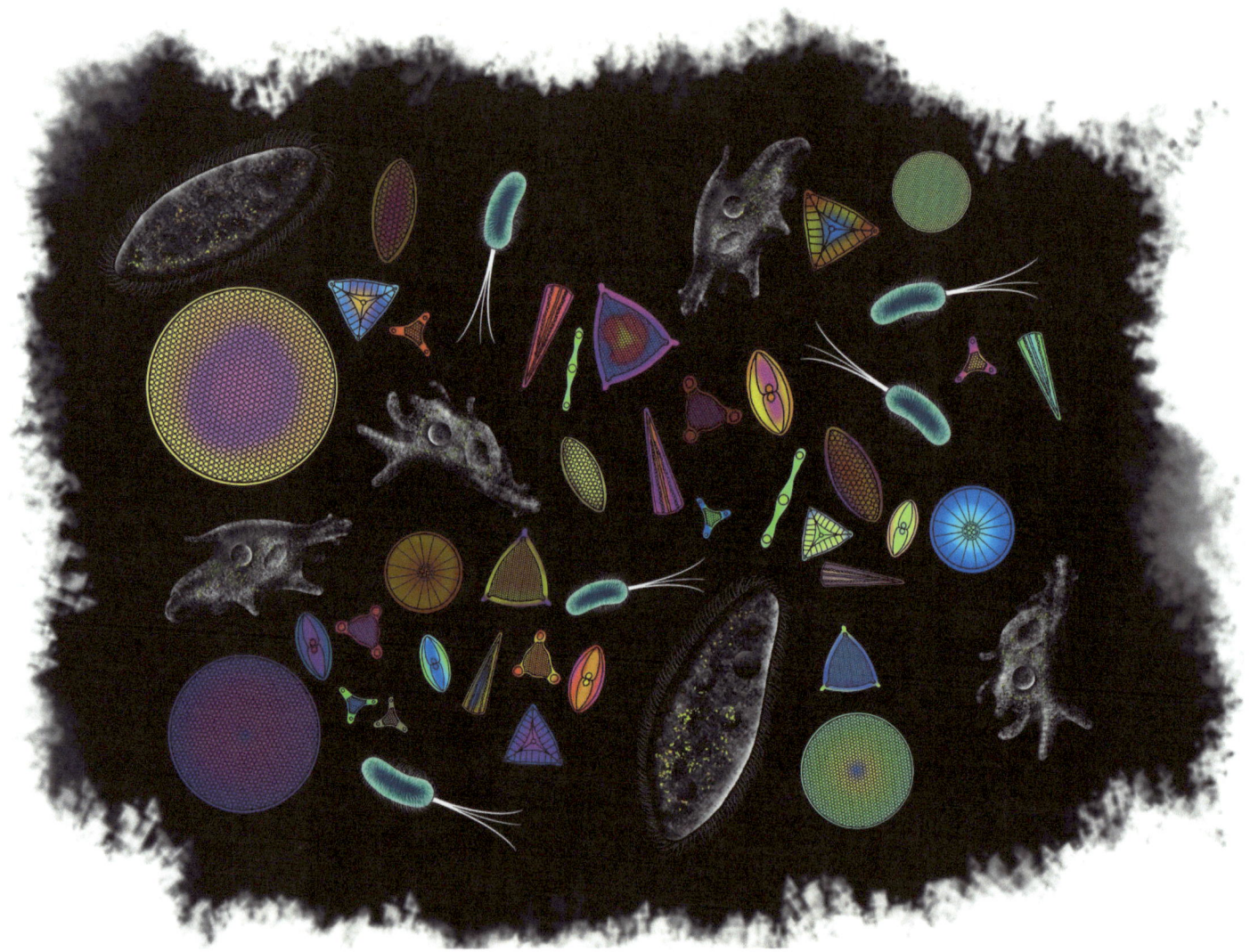

and are called unicellular

Some are made of many, many cells

and are called multicellular

You and I are made of

many different types of cells

Our cells are what

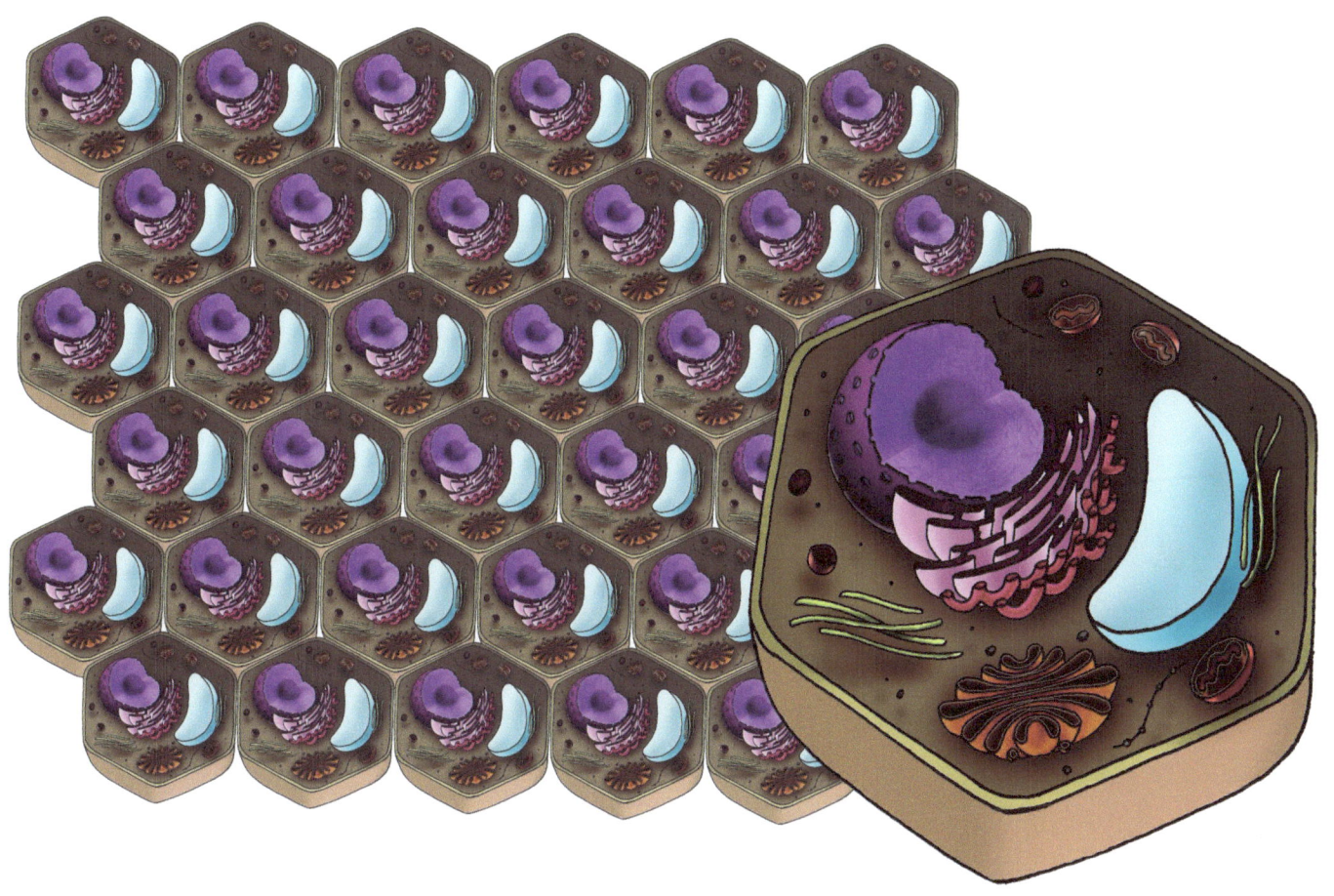

that keep us going

They help us do

what we are doing

You move

your hands and legs

With your mighty

muscle cells

You think beautiful

thoughts of love

You get fresh

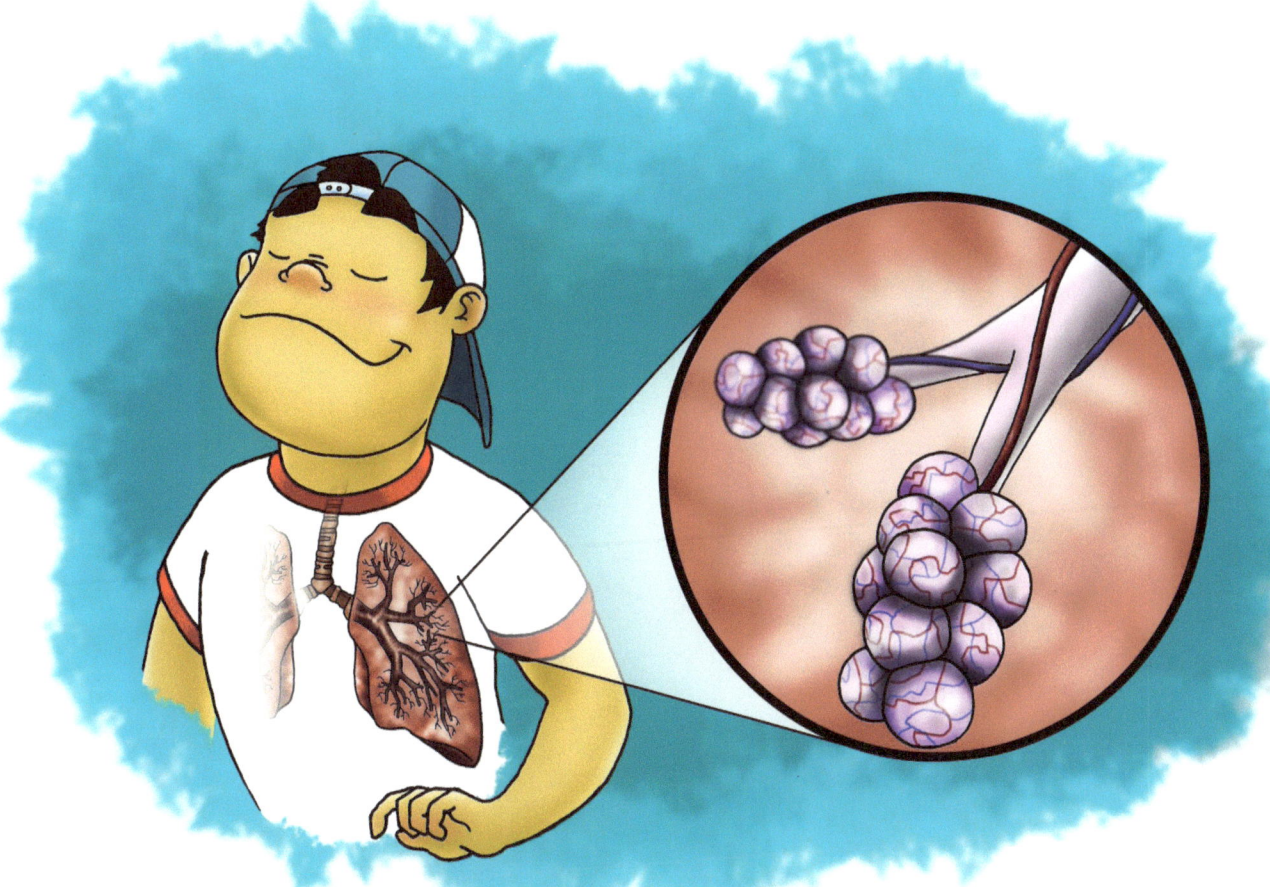

Oxygen to breathe

Through your

red blood cells

You see the rainbow

and other pleasant sights

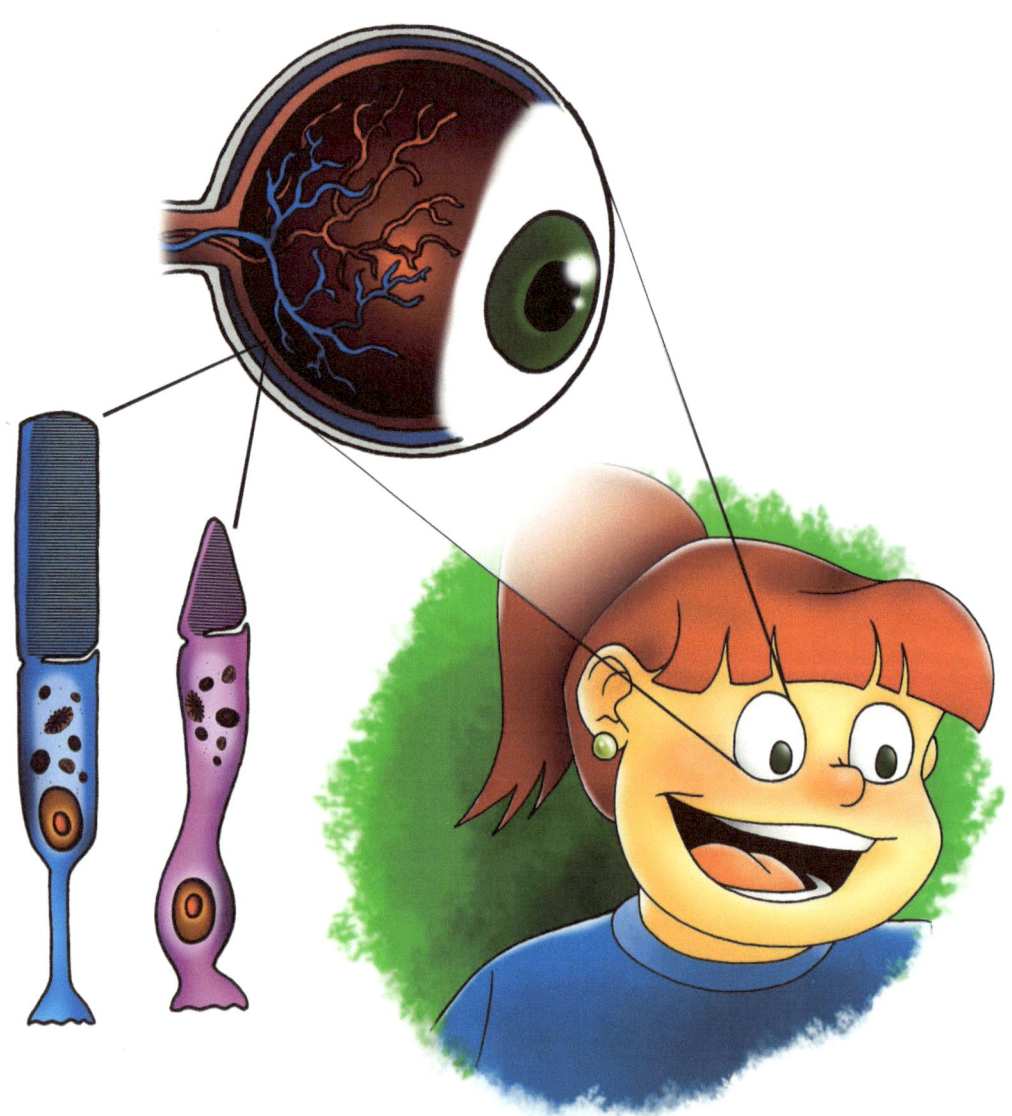

We can keep on going

As there are so many other cells

But now you know

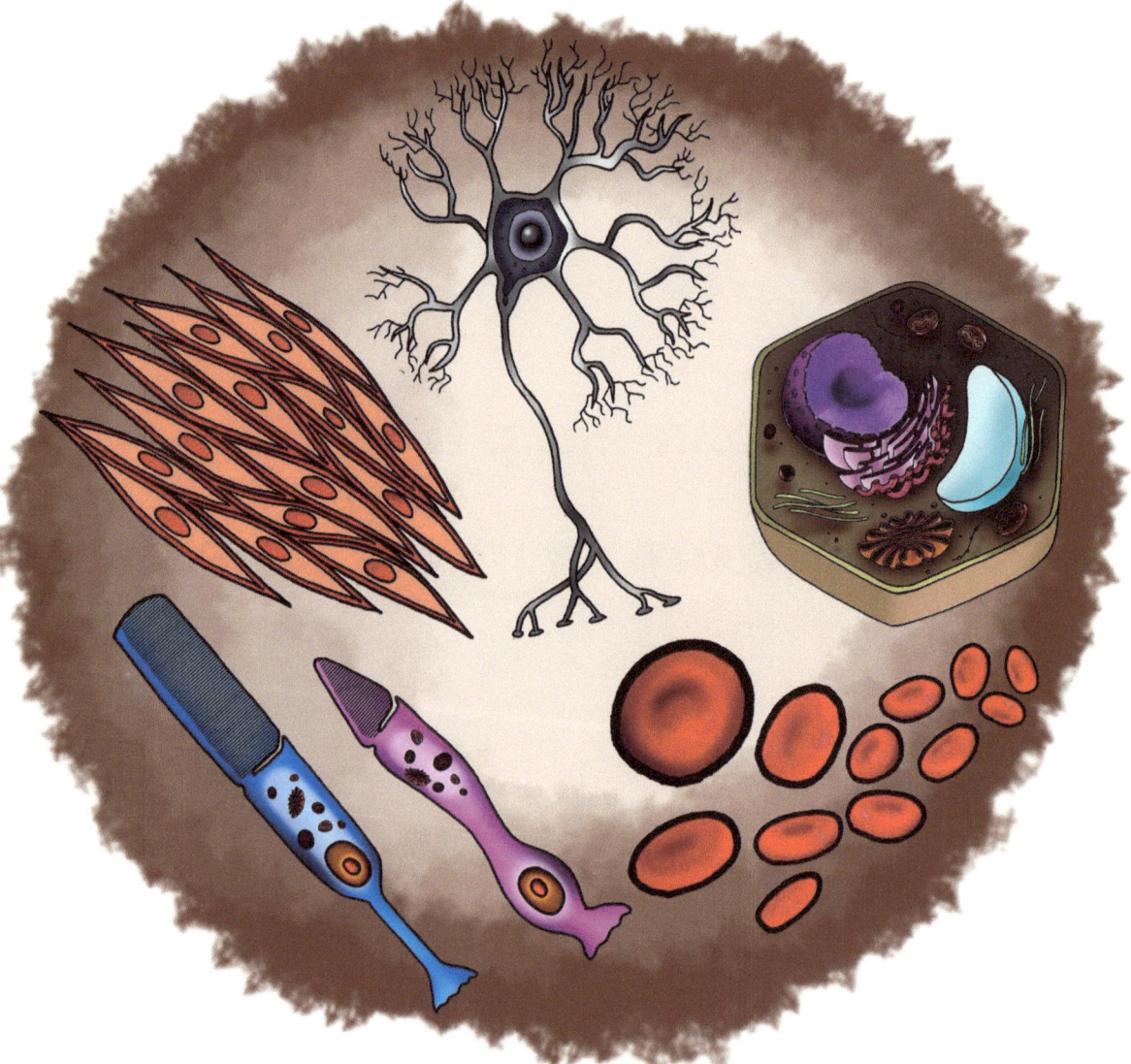

You and I are made of many special cells

We can keep on going

As there are so many other cells

But now you know

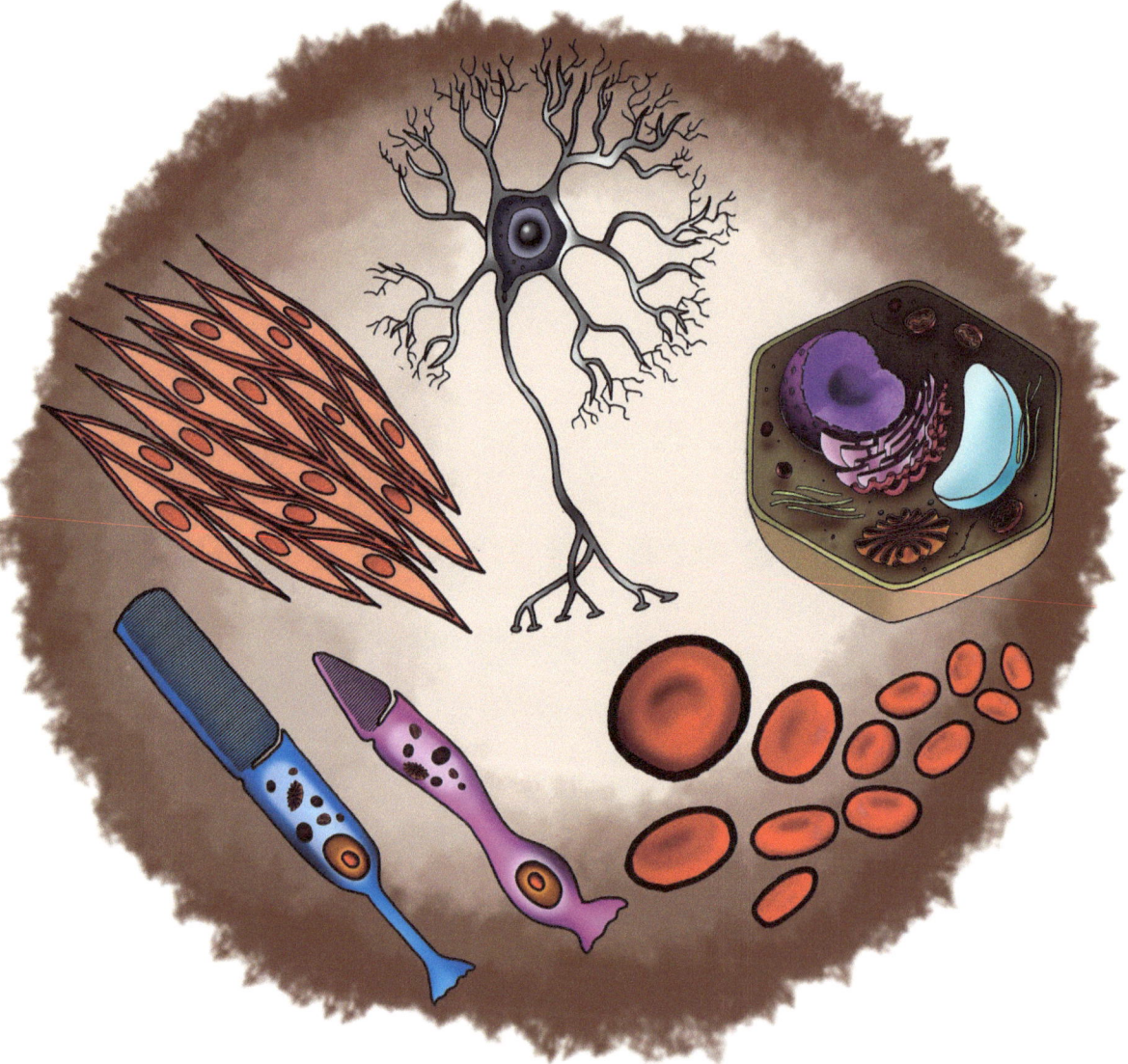

You and I are made of many special cells

FREE BONUS – Cells Coloring Pages

As a thank you for buying this book, you will find the link to get printable free cell coloring pages at
http://freecellscolor.weebly.com

Bedtime Science Series
Making Scientific concepts accessible EARLY and EVERYWHERE

Coming Soon:
- Inside a Cell
- Cellular Highway
- Into the Brainy Brain
- All in a day's work

Copyright © 2016 Ela Joseph
All rights reserved.

www.ingramcontent.com/pod-product-compliance
Lightning Source LLC
Chambersburg PA
CBHW041308180526
45172CB00003B/1026